Birds of Culpeper County

in the

Virginia Piedmont

Lee Alloway

Ancient Eagle Press
Culpeper, Virginia
2025

ISBN: 978-0-9971923-4-6

Library of Congress Control Number: 2024903156

ACKNOWLEDGMENTS

I am not an ornithologist. For the past two decades I have been a photographer of bugs. However, where there are bugs there are birds, and it didn't take long to fall in love with those as well. The jungles and mountains of the world host a great number of colorful exotic birds, but our local species have a beauty of their own. I've tried to capture that in pictures I have taken of birds in Culpeper County...birds that are commonly found throughout the Virginia Piedmont.

This book would not have been possible without the patience and assistance of my wife, who is the birder in the family. Jan introduced me to birds thirty years ago and knows more about them than I ever will.

Table of Contents

Introduction

The Virginia Piedmont, stretching from the Blue Ridge mountains to the Atlantic coast, supports an impressive variety of native bird species and is seasonally visited by many migratory species heading north to breed or south to winter in the sunshine. Culpeper County, with its proximity to mountains, lakes, and agricultural areas, provides suitable habitat for diverse species, offering many opportunities to observe and enjoy the birds while they share their time with us.

This volume grew from a previous project that documented 1,000 species of local bugs. While bugs are fascinating and technically challenging to photograph, they rarely raise the spirit or inspire song as birds do. Birds fill our days with music from morning reveille to evening vespers. They accent our sky and trees with every color of the rainbow. They entertain us at the feeders and reciprocate our generosity by removing bugs from our gardens. Some live with us year-round while others simply stop by during migration, but all remind us of the diversity and beauty of nature around us. This book is my tribute to those birds[1]. Photos follow, but first:

A Bit About Birds

How many birds are alive and flying today? Nobody knows for sure. In a 2019 study, biologist Corey Callaghan and his team at the University of New South Wales estimated the population of the 9700 most populous species (cited as 92% of all bird species) to be about 50 billion, noting that inclusion of 1200 less populous species and the application of other assumptions in their model could indicate a total bird population of 428 billion[2]. Talk about hedging your bets! Earlier estimates have posited 100-300 billion. It's safe to say there are LOTS of birds.

Consistent with the widely varied estimates of bird populations, there are a variety of opinions about how many species exist. In 2022, the Clements Checklist of Birds of the World listed 10,906 bird species in 249 families and 41 orders. That is an incremental increase over the 9,797 species listed 20 years earlier by Birdlife International, and includes new species discovered or split from existing families.

---

[1] There is a high likelihood you will see birds not mentioned in this book. In the end of the book you will find several excellent references to help you identify those.

[2] Callaghan, C.T. et al, *Global abundance estimates for 9,700 bird species,* PNAS 2021 vol.118 No.21

Other taxonomists, however, contend there are almost 13,000 species based on genomic diversity. More than half of all species are found in South America or Australia. The U.S. has about 1,200 species[3].

> The most common bird in the world is the domestic chicken, with 25 billion of various varieties. Among wild birds, the top three are all represented in our area.
>
> - House Sparrow (1.6 billion worldwide)
> - European starling (1.3 billion worldwide)
> - Ring-billed Gull (1.2 billion worldwide)

Classification

For thousands of years people have been studying and classifying animals. The best known system of classification (taxonomy) is the binomial nomenclature devised by Carl Linnaeus in the mid-1700s[4]. While Linnaeus' system describes three Kingdoms of nature (Plant, Animal, and Mineral), we will only be concerned with animals.

Within the animal Kingdom, individuals are grouped into

- Phyla (singular "Phylum") which contain
  - Orders (singular "Order") which contain
    - Families (singular "Family") which contain
      - Genera (singular "Genus") which contain
        - Species (singular "Species").

The "binomial nomenclature" refers to the Genus and Species of an individual. A Northern Cardinal for instance is in the Animal kingdom, Phylum *Chordata*, Class Aves[5], Order *Passeriformes*, Family *Cardinalidae*, Genus *Cardinalis*, and Species *cardinalis,* so its binomial nomenclature is *"Cardinalis cardinalis."*

While Linnaeus got things started, there are now more than one million species that have been classified, including 10,000-13,000 birds. To prevent

[3] eBird currently lists 1151 species reported in the United States, including vagrants that are not considered residents.

[4] The best known early taxonomy is *Historia Animalium* written by Aristotle in 350 BC

[5] Birds

overcrowding, taxonomists have created numerous subdivisions such as "subphyla," "superorder," "suborder", "subfamily," etc., but the basic structure remains intact. The system has proven to be flexible enough to accommodate renaming and shifting of individuals within the system as more is understood about the true nature and relationships of the individuals. Increased use of genetic sequencing has led to frequent reclassifications. Classifications used in the volume are current as of 2023 but are all subject to revision as scientists learn more.

How did they get here?

A credible answer is, "They flew." But let's look a little farther back. Birds are the current evolutionary stage of prehistoric reptiles: the dinosaurs. For almost a century, a single feathered dinosaur—*Archaeopteryx macrura* — was considered the predecessor of all birds. Discovered in Germany in 1861, a fossil of *Archaeopteryx* from the Late Jurassic (about 150 million years ago) features feathers and distinct dinosaur skeletal structures. More recent discoveries indicate even older precursors of modern birds, suggesting several evolutionary lines.

While the fossil record shows a gradual development of feathers from skin cells, a more vexing question is how cold-blooded reptiles became warm-blooded birds. Recent studies have shown that a large lizard, the Argentine black and white tegu (*Salvator merianae*), can maintain its body temperature up to 18 degrees above its environmental temperature in the winter.[6] Since these modern reptiles have the capability to thermally regulate, it is less surprising that full warm-bloodedness has evolved in birds.

While taxonomists may debate the proper relationships and lineage of bird species, there is general agreement that the fossil record and DNA analysis support the dinosaur as precursor to birds. There is also the hoatzin (*Opisthocomus hoazin*).

Hoatzin
(Opisthocomus hoazin)
Not found in Culpeper!

[6] Certain other lizards as well as pythons are also partially endothermic.

This South American swamp dweller, about the size of a pheasant, looks like it would be right at home in Jurassic Park. The chicks of this bird have claws on their outer two wing digits, similar to *Archaeopteryx* (and a few other modern birds). The adults eat leaves which they chew and digest by fermenting them in their crop (foregut fermentation) like cattle. Taxonomists have placed the hoatzin in its own Class. It is a tribute to its dinosaur past!

Physiology

Despite being the modern form of flying lizards, birds would seem to have much in common with us. They are warm-blooded chordates with two eyes, a nose, mouth, ears, two legs, and two (feathered) arms with modified hands. However, many of those parts work in strange and wonderful ways to support a variety of survival strategies.

Humans have binocular vision brought to us by a pair of steerable eyes that track together and have detailed focal area in the center of the eye with less precise peripheral vision. Bird eyes are not steerable and are not oriented in the same direction[7], and birds often lack the visual acuity of humans.[8]

There are some notable cases of extreme visual acuity in birds, particularly among the raptors. Eagles can spot a small rodent from a mile away if there is good light. Eagle eyes as well as those of vultures are modified to have two separate focal areas, one straight ahead and one offset by 45 degrees that offers higher resolution. That arrangement leaves a blind spot above and below the bird. For black vultures, the double focal area arrangement allows them to scan for food below while staying oriented to other vultures in their kettle. The downside for vultures as well as eagles is that while looking down for prey, the blind spot is directly in front of the bird, which is why a bird that can see a 6" rodent a mile away will fly into a 300' wind turbine directly in front of it.

Other birds have other requirements of their eyes. A mallard's eyes take in a panorama, keeping him aware of everything around him, threats as well as

---

[7] The exception is the owls, whose eyes are both oriented forward in fixed sockets but can be 'steered" by rotating the head. Our local barred owl can rotate its head 270 degrees in either direction. They can look right by turning their head left!

[8] Humans can distinguish objects separated by 1/60th of a degree of arc, while half of all birds can only discriminate 1/6th of a degree...a standard at which a human is considered legally blind.

opportunities. He sees where he's going as well as where he's been, plus everything above him. A heron can look straight ahead and see fish at its feet.

Birds may also see at a different speed than humans. We can distinguish and react to visual cues at the rate of 60 events per second. Birds such as flycatchers who depend on being able to catch insects in flight may be able to discriminate visual cues at twice that speed.

As humans, we see a spectrum of colors by interpreting input from sensors that detect red, green, and blue. Our brain compares the incoming data and reports the result as a single color. Many birds have a fourth sensor, one that detects ultraviolet light. While it is impossible to determine what "color" the bird's brain reports, it serves at least one very important purpose. Some species in which male and female look alike to us, such as our Northern Mockingbird, have feathers that reflect different UV patterns in male and female. As clearly as we can see the difference between a male and female cardinal, birds with ultraviolet sensors can see the difference between male and female mockingbirds.

Hearing is another area where birds show extraordinary adaptations. They hear faster than we do. We can distinguish sounds that are separated by as little as 4 milliseconds. Songbirds can detect separations of 1-2 milliseconds (varying by species). When we hear the song of a bird, what we perceive as a single note may be the average of several different notes produced by the bird, notes that convey meaning to other birds but are lost on us. At different times of year, birds' hearing changes. In the spring, hearing is tuned to hear the subtleties of mating calls; in the fall, it is more sensitive to other sounds that signal danger or migratory cues from their fellow songbirds.

Speaking of hearing, some owls have another adaptation. Being night hunters, they rely on their hearing to find prey, using their relatively good low light vision as necessary only at close range. Their secret weapon is that their ears, located behind their eyes, are at different levels, enabling them to determine not only the direction of the sound, but also the elevation. We hear in stereo. They hear in 3D and can pinpoint their prey in total darkness, even if it is hidden under snow.

In addition to sensory adaptation, some birds can change their physical structure. Case in point: the gizzard. When a bird eats, food is stored in the crop, then passed to the stomach where it is digested. Food that cannot be digested in the stomach goes back the way it came in or is passed to the gizzard where strong muscles mash and grind the food into digestible pieces. If the bird's food source provides easily digestible food (worms, caterpillars, fruit) the gizzard has little work to do, so it shrinks to a very small size. As food sources change to include harder to digest things like nuts and seeds, the gizzard grows. After putting songbirds on a high-fiber diet, researchers found the gizzard began growing within 24 hours.

Just as the gizzard is modified to accommodate the diet, the reproductive system undergoes transformation during the breeding season. Ovaries, almost imperceptible most of the year, grow rapidly to accommodate egg production, then shrink dramatically when breeding season is over.

In most birds, only the left ovary and oviduct are functional. Although the female embryo has two ovaries, the right one typically regresses and is nonfunctional in the adult bird.

Why are they in Culpeper?

From an environmental perspective, Culpeper has much to offer.

- We are far enough south that we don't have to endure the brutal New England winters.
- We are far enough north that we escape searing summer heat.
- When hurricanes rake the East Coast, we are far enough inland to be safe from the storm surge and flooding.
- And the wicked weather of the Midwest loses some of its fury when it hits the Blue Ridge mountains.

Yes, we have great weather, but why are these specific birds in Culpeper rather than Ashville or Pittsburgh?

Much can be attributed to genetics. Species have evolved to live where there is reliable food and shelter. For some, that means they will stay year round. Others begin their migrations in response to environmental triggers, which may be the length of day, changes in weather patterns, or changes in food availability. But how do they find their way here? How do they even know where "here" is?

Magnets! We are taught in school that we have five senses: sight, hearing, taste, touch, and smell. That's a good start, but in fact we have many more...as many as 20 senses have been identified, including vestibular sense and proprioception. Migratory birds have at least one that we don't have. They have magnetic sensors. When they are born, many birds have the magnetic signature of their location imprinted in their brain. When they migrate, they are guided by the earth's magnetic field as well as visual and olfactory cues learned on previous migrations. Severe weather may cause them to deviate from their migratory path and changes in their home range might cause them to move their territory, but generally birds will travel predictable migratory routes and maintain consistent summer and winter territories. Some will travel thousands of miles and return to the same nest sites year after year. Although the phenomenon of magnetoreception is well documented, the location and nature of the receptors themselves is still undiscovered.[9]

What are they up to?

The end of winter may be seen as the beginning of the life cycle of our birds as they establish territory, compete for a mate, and raise a family, all timed to the emergence of appropriate food.

When birds arrive in the spring, they need a home and a good meal.

- Bushes, trees, and marsh grass are traditional nesting sites, but birds often take advantage of man-made shelters. That can be bird houses, but anything that provides protection from weather and predators will do.
- Although some birds will eat flowers and grasses, the real advantage of having shoots and buds is that bugs emerge as the plants do in the spring, and it is the bugs that the birds love to eat; bugs are essential for raising chicks.

Why are bugs necessary? Birds have no teeth[10]. As previously noted, food that has to be ground up is processed in the gizzard, but the gizzard doesn't work on

---

[9] A leading theory is that two molecules in the eye, when triggered by light, create paired electrons that are sensitive to magnetic fields.

[10] The bills of some aquatic birds such as dabbling ducks have serrated ridges that filter food from the water, but those are not properly teeth. Similarly, an egg tooth, present on many chicks to facilitate their escape from the egg, is not a tooth but a temporary projection of the bill.

muscle power alone. After fledging, birds eat grit and small stones that travel to the gizzard where they are the mechanical part of the grinding process. Chicks, having never left the nest, have no access to grit, so have no ability to process hard foods. Worms, caterpillars, suet, and fruit are excellent chick food. Adults that bring insects with hard exoskeletons back to the nest may trim off elytra and mandibles before offering the remaining digestible parts to the chicks.

Incubation and fledging may take six weeks or longer. Some species continue to nurture chicks for extended periods, while others may have second or even third broods over the summer. After mating season, many birds molt, exchanging worn feathers for a new set. The new feathers will gradually wear down over the following months, in many cases showing their most vibrant colors just in time for the next mating season.

Is That Bob?

One of the wonderful things about spring is the return of migrating birds. The warblers, grosbeaks, and orioles that graced your feeders have returned. Or have they? Is that really "Bob", the same bird that has eaten your safflower seed for the past five years? Has "Martha" found her way back from vacationing in Florida? Probably not.

Birds have a hard life, and often a short one. Island and ocean-dwelling birds tend to live longest, raptors fare well, but for migrating songbirds, life is a struggle. As a general rule, small birds live shorter lives than large birds, and songbirds are at the top of the mortality tables. For some, there is only a 25% chance of surviving to adulthood, and about a 50/50 chance of surviving the first migration cycle. So chances are "Bob" has been replaced by "Bennie" at your feeder after a year or two.

Why the high mortality rate? By far, the largest factor is domestic cats! In the U.S. alone, cats kill more than 500 million birds each year, more than all other causes combined. Collisions with cars, buildings, and power lines cause another 125 million deaths, with disease and habitat loss adding to the toll. These factors have combined to drive down the migratory bird population by 28% over the past five decades. The U.S. bird population has decreased by 2.9 billion individuals, with more than half of the grassland birds now gone. Most of those losses come from 12 families, including sparrows and warblers.[11] An exception to the decline is the

---

[11] Rosenberg, K.V. et al, Decline of the North American Avifauna, *Science*, AAAS, 2019

waterfowl that have increased by more than 50% over the same time thanks to wetland preservation and adaptive farming practices.[12]

How can we help?

Our local birds and bugs have evolved with our local plants. As we put our imprint on our yards, we should consider the birds and bugs, and plant accordingly. I know we all love flowers, and there is certainly joy to be found in a well-tended garden of colorful flowers, many of which are not native. There's nothing wrong with that in moderation if there are suitable plants around that will support all stages of the bugs.

Case in point: The butterfly bush. Gardeners love them because they are hardy, require little maintenance, and have abundant flowers that attract butterflies and bees. However, they are not native, and no local bugs eat them. Many insects, however, lay their eggs where they feed. If a butterfly lays its eggs on a butterfly bush, the eggs will hatch but the caterpillars will starve to death. So, if you have a butterfly bush in your yard, be sure there is a reliable caterpillar food source nearby where the butterflies can lay their eggs so the birds will be well fed.

To nurture birdlife, trade at least a portion of your lawn for native trees and bushes to provide food and nest sites. Minimize your use of pesticides. Supplemental feeding (i.e., bird feeders) will bring more birds into your yard and benefit the birds during periods of drought or severe cold. In the spring, meal worms, fruit, and suet will benefit chicks. Periodic cleaning of feeders is important to minimize transmission of disease.[13]

Minimize outdoor lighting not needed for safety and security. Artificial lighting disrupts insect breeding patterns and results in disrupted food supplies for the birds. Artificial light also disorients birds during migration.

Above all, keep your cats inside.

---

[12] Per the Audubon Bird Count, crows are up 250% in the same time...a mixed blessing at best.

[13] Soap and water is generally sufficient, although an occasional rinse with a mild bleach solution will minimize transmission of diseases such as avian conjunctivitis.

Birding in Culpeper

If you fill a feeder in your yard, within days the birds will discover it. Finches, cardinals, chickadees, and nuthatches will be regular visitors, joined by occasional woodpeckers and others. Some birds are not attracted to feeders but may be found at any of the Culpeper birding hot spots.

- Lake Culpeper is a prime location for waterfowl, warblers, and sparrows.
  - Lake Culpeper Adventures by the Olde Country Store provides easy access and will rent canoes and paddleboats. Traveling along the shoreline may reveal a wide variety of birds. One hundred species included in this book were photographed on or near Lake Culpeper.
  - The former boat ramp at Henretty Drive
  - Areas around Lake Culpeper dam
- Mountain Run Lake Park
- Yowell Meadow Park
- Culpeper War Memorial Park
- Culpeper Sports Complex
- For vultures, the solid waste transfer station (14017 Laurel Valley Road, off Rt 522)

Most birds are active all day but activity peaks in the morning as birds are out for their morning nourishment and singing to claim their territory. Another uptick occurs in the evening as flocks return from foraging. Best birding tends to be the two hours after sunrise and before sunset.

Resources/Supplies

Wild Birds Unlimited (15311 Montanus Dr) – Information, activities, books; bird food/feeders, bird houses, and related products

Tractor Supply (588 Culpeper Town Sq) – Good supply of bird food, including bulk quantities.

Now let's go birding

Order Anseriformes -- Ducks, Geese, and Swans

Family Anatidae (Geese, Ducks, Swans)

Wood Duck
(Aix sponsa)

Length	19-21"
Wingspan	26=29"
Weight	16-30 oz.

The wood duck is the only native perching duck found north of the Rio Grande. It prefers to perch high in the trees and may nest in tree cavities. It tends to have large broods but there is high infant mortality.

Dabbling Ducks

Green-winged Teal
(Anas carolinensis)

Length	12-15"
Wingspan	20-23
Weight	**5-17 oz.**

Teal, mallards, and wigeons are dabbling ducks. With legs in the middle of their undercarriage they easily turn upside down to acquire food in shallow water.

Mallard
(Anas *platyrhynchos*)

Length	20-26"
Wingspan	32-39"
Weight	**1.5-3.5 lb.**

American Wigeon
(*Mareca americana)*

Length	17-23"
Wingspan	30-36"
Weight	1-3 lb.

Diving Ducks

Lesser Scaup
(*Aythya affinis*)

Length	15-19”
Wingspan	27-31”
Weight	1-2.5 lb

Ring-necked Duck
(*Aythya collaris*)

Length	15-18”
Wingspan	24-25”
Weight	1-2 lb.

Length 16-17”

21-23”

Ruddy Duck
(***Oxyura jamaicensis***)

Length	13-17”
Wingspan	18”
Weight	1.25 lb.

Diving ducks (subfamily Aythyinae) dive under water in search of snails and insects. Their legs are located at the aft end of the undercarriage to facilitate swimming under water. They usually swim low in the water, sometimes only the head and neck are visible.

Sea Ducks

Bufflehead

(Bucephala albeola)

Length	13-16"
Wingspan	22"
Weight	10-20 oz.

Buffleheads have evolved their small size to fit the nesting cavity of their "metabiotic" host, the northern flicker. Buffleheads are monogamous and the females may return to the same nest site, year after year.

Sea ducks are relatively long-lived in the wild (up to 7 years) but have small broods with low infant mortality. Dabbling ducks like the mallard have large broods but short lives (2 year average) and high infant mortality.

Common merganser

(Mergus merganser)

Length	**23-28"**
Wingspan	**30-38"**
Weight	**2-4.5 lb.**

Canada Goose

(*Branta canadensis*)

Length	30-43"
Wingspan	50-73"
Weight	5-14 lb.

Nobody's favorite bird, the Canada goose is large, loud, and abundant. Protected by the Migratory Bird Act, scores of Canada geese are permanent residents of Lake Culpeper, joined by hundreds of their kind in the fall and winter.

Mute Swan

(*Cygnus olor*)

Length	55-63"
Wingspan	79-94"
Weight	20-32 lb.

Our most abundant swan, recognized by its orange bill. This is an introduced species.

Trumpeter Swan

(*Cygnus buccinator)*

Length 54-65”
Wingspan 74-122”
Weight 15-30 lb.

The trumpeter swan is the largest extant species of waterfowl, and both the heaviest and longest native bird of North America. Juveniles are grey.

Tundra Swan

(*Cygnus columbianus*)

Length 45-59”
Wingspan 66-83”
Weight **7**-21 lb..

Somewhat smaller than the trumpeter swan but looking very much the same, a Tundra swan (aka Whistling swan) is seen here in company of two adult and two juvenile Trumpeter swans.

The largest modern bird on record was the wingless South Island giant moa (*Dinornis robustus*) which stood 12' tall (although bone structure suggests it would customarily carry its neck outstretched for grazing rather than upright), weighing 530 lbs. The dominant animal in New Zealand for millions of years, it became extinct due to hunting and habitat loss within 100-years after the Māori arrived around 1300 A D.

The tallest living bird is the common ostrich (Struthio camelus), standing 7 to 9 feet tall. At 200-290 lbs., it is also the heaviest.

The smallest known bird is the bee hummingbird (Mellisuga helenae), a Cuban endemic less than 2 ¼ inches long and 0.08 oz. With a wingbeat of up to 200 beats per minute and a heart rate of up to 1250 beats per minute, the bee hummingbird may live up to seven years eating nectar in the wild but is listed as Near Threatened due to habitat loss.

Order Charadriiformes -- Sandpipers and Seagulls

Family Laridae (Gulls, Terns)

About 100 species of seabirds are in the Laridae family, with three frequently found in the Culpeper area.

Killdeer

(Charadrius vociferus)

Length	8-11"
Wingspan	23-25
Weight	2.5-4.3 oz.

This common resident of Lake Culpeper's shallows and shore can occasionally be found in nearby grassy areas. Less common in the winter.

Bonaparte's Gull

(*Chroicocephalus Philadelphia*)

Length	11-15”
Wingspan	30-33”
Weight	6-9 oz.

Ring-billed Gull

(*Larus delawarensis*)

Length	17-21”
Wingspan	41-46”
Weight	11-25 oz.

Family Scolopacidae (Sandpipers and Snipes)

Almost 100 species of shorebirds are in the Scolopacidae family. Most are small birds that feed on invertebrates they find by probing the mud with their long bills. Variations of bill length and shape accommodate variations in diet and allow several species to coexist with minimal competition for food. Six species are commonly or seasonally found in Lake Culpeper.

Spotted Sandpiper

(*Actitis macularius*)

Length	7-8"
Wingspan	15-16"
Weight	1-2 oz.

Pectoral Sandpiper

(*Calidris melanotos*)

Length	8-10"
Wingspan	18"
Weight	2.6 oz.

Least Sandpiper

(*Calidris minutilla*)

Length	5-10"
Wingspan	11"

Solitary Sandpiper
(Tringa solitaria)

Length	7-9”
Wingspan	20”
Weight	1-2.3 oz.

Lesser Yellowlegs
(Tringa flavipes)

Length	9-10”
Wingspan	23-25”
Weight	2.4-3.3 oz.

Wilson's Snipe

(Gallinago delicata)

Length	8-11"
Wingspan	15-18"
Weight	3-5 oz.

If you're ever sent on a snipe hunt, this is what you're looking for.

Order Pelecaniformes – Herons, Spoonbills

Family Ardeidae (Herons and Egrets)

Great Blue Heron

Ardea herodius

Length	36-54”
Wingspan	66-79”
Weight	4-8 lb.

Numerous herons spend their summers in Culpeper, but most migrate south in the winter. Our few cold weather herons have likely migrated here from breeding areas up north.

Green Heron

Butorides virescens

Length	16-18"
Wingspan	25-27"
Weight	8-9 oz.

Juvenile

Black-crowned Night Heron

(*Nycticorax nycticorax)*

Length	23-26"
Wingspan	45-46"
Weight	26-36 oz.

Great Egret

(Ardea alba)

Length	31-41"
Wingspan	52-67"
Weight	1.5-3.5 lb.

Family Threskiornithidae (Spoonbills and Ibises)

Roseate Spoonbill

(Platalea ajaja)

Length	28-34"
Wingspan	47-52"
Weight	2.5-4 lb.

The roseate spoonbill is associated with Florida but for the past few years climate change and habitat loss have caused individuals to fly north for the summer. These were in Lake Culpeper in the summer of 2022 and 2023.

Order Suliformes -- Cormorants

Family Phalacrocoracidae (Cormorants)

Double-crested Cormorant

(Nannopterum auritum)

Length	28-35"
Wingspan	24-48"
Weight	3-6 lb.

A capable swimmer on the surface and under water, the cormorant is primarily a fish eater. One of the few aquatic birds whose feathers are not waterproof, the cormorant spends considerable time drying its wings.

Order Coraciiformes -- Kingfishers

Family Alcedinidae (Kingfishers)

Belted Kingfisher
Ceryle alcyon

Length	11-14"
Wingspan	19-23"
Weight	4-6 oz.

Several pairs of kingfishers maintain territory around Lake Culpeper in the summer and a few individuals have been present in the winter. In a case of "reverse sexual dimorphism", the female kingfisher sporting a chestnut band is more colorful than the male.

Order Gruiformes -- Cranes, Rails, and Allies

Family Rallidae

American Coot
(Fulica americana)

Length	13-17"
Wingspan	23-28"
Weight	1-2 lb

Also called a Mud Hen (the namesake of the Toledo minor league baseball team), coots are occasional visitor during their migration. Their large feet are adapted for walking in the shallows and on land.

Order Accipitriformes -- Hawks, Eagles, and Vultures

Family Accipitridae (Hawks, Eagles)

Bald Eagle

(*Haliaeetus leucocephalus*)

Length	28-40"
Wingspan	71-91"
Weight	6-14 lb

Our national bird, the bald eagle is a common resident of the Culpeper area (less so in the winter). With a lifespan of 20 years in the wild, it takes four or five seasons before the juveniles reach sexual maturity and take on their adult plumage. Although best known for catching fish, the eagle will catch other large birds or small mammals, and will frequently feast on carrion, particularly when the lake is frozen.

i

Red-shouldered Hawk

(*Buteo lineatus*)

Length	15-24"
Wingspan	35-50"
Weight	1-2 lb

Our most common hawk, found throughout the year. Diet is varied, including small mammals, frogs, snakes, fish, and carrion. It will also take birds from your bird feeder.

Red-tailed Hawk

(*Buteo jamaicensis*))

Length	18-26”
Wingspan	43-56”
Weight	1.5-3.5 lb

Our largest local hawk, the red-tail is a consumer of rabbits, rats, and squirrels, as well as the occasional bird. Females are about 25% larger than males.

Cooper's Hawk

(Accipiter cooperi)

Length	14-18"
Wingspan	24-39"
Weight	0.75-1.25 lb

Formerly known as the chicken hawk for its preference for medium-sized fowl, the Cooper's hawk may again be renamed as taxonomists of the American Ornithological Society plan to rename birds currently named after persons. An opportunistic predator around bird feeders.

Family Cathartidae (New World Vultures)

Turkey Vulture
(*Cathartes aura*)

Length	24-32"
Wingspan	63-72"
Weight	2-5 lb

Black Vulture
(*Coragyps atratus*)

Length	22-29"
Wingspan	52-66"
Weight	3.5-6.5 lb

Both vultures are common in our area and often congregate in areas of open dumpsters. Black vultures can reliably be observed at the solid waste transfer station on 522. Turkey vultures have elaborate channels in their noses that enable them to find food from far away. Black vultures are less nasally blessed but have excellent eyesight and follow turkey vultures to dinner.

Family Pandionidae (Osprey)

Osprey

(*Pandion haliaetus*)

Length	24-32”
Wingspan	50-71”
Weight	2-4.5 lb

Groups of birds are called various things more descriptive than "flock." Many of the terms are regional, and the same birds may have different names in flight than when on the ground or water. You probably know about a gaggle of geese and perhaps a murder of crows, but here are a few others to ponder:

An affliction of starlings

An annoyance of house sparrows

An aristocracy of waxwings

A bubbling of wrens

A congregation of cardinals

A creep of nuthatches

A farce of jays

A gang of turkeys

A Gatling of woodpeckers

A kettle of hawks

A Parliament of owls

A rainbow of grosbeaks

A stealth of cowbirds

A yellowing of warblers

Order Strigiformes — Owls

Barred Owl

(*Strix varia*)

Length	16-25"
Wingspan	38-49"
Weight	1-2 lb..

Rarely seen but frequently heard, the barred owl serenades at night with its call whose rhythm calls to mind "Who cooks for you, who cooks for you-all".

Eastern Screech Owl

(*Megascops asio*)

Length	6-10"
Wingspan	18-24"
Weight	6-7 oz.

The screecher's ear tufts are strictly decorative. The actual ears are behind the eyes and slightly off set to enable locating prey.

Order Galliformes -- Grouse, Phesants, Turkeys

Family Phasianidae (Chickens, Turkeys, Pheasant)

Wild Turkey

(Meleagris gallopavo)

Length	39-49”
Wingspan	49-57”
Weight	11-24 lb

Although Ben Franklin may not have really supported the turkey as America’s national bird, he did refer to it as a “...respectable bird, and withal a true original native of America.” The toms can be quite large and in protected areas can be aggressive toward humans.

Order Piciformes--Woodpeckers

Family Picidae (Woodpeckers and Sapsuckers)

Northern Flicker

(*Colaptes auratus*)

Length	11-14"
Wingspan	16*21"
Weight	**3-6** oz.

Yellow-bellied Sapsucker

(*Sphyrapicus varius*)

Length	7-8"
Wingspan	13-16"
Weight	1-2 oz.

Red-headed Woodpecker

(*Melanerpes erythrocephalus*)

Length	6-10"
Wingspan	16-17"
Weight	2-3.5 oz.

Previously listed as Near Threatened due to decreasing populations attributed to habitat loss, the red-headed woodpecker population seems to have stabilized at its lower level. This bird nests in dead trees. Clearing dead trees from your acreage is removing habitat vital to bugs and birds.

Pileated Woodpecker
(*Dryocopus pileatus*)

Length	16-19”
Wingspan	26-30”
Weight	8-14 oz.

Red-bellied Woodpecker
(*Melanerpes carolinus*)

Length	9-11”
Wingspan	15-18”
Weight	2-3 oz.

Hairy Woodpecker
(*Picoides villosus*)

Length	7-10"
Wingspan	13-17"
Weight	1.5-3.5 oz.

Hairy and Downy have similar patterns, but Harry is larger with a relatively longer beak. Females of both species lack the red patch on the back of the head. The downy is much more common in our area and a year-round resident.

Downy Woodpecker
(*Picoides pubescens*)

Length	5.5-7
Wingspan	10-12"
Weight	0.7-1.2 oz.

Order Apodiformes[14]-- Hummingbirds

Family Archilochus

Ruby-throated Hummingbird
(*Archilochus colubris*)

Length	3-3.5"
Wingspan	3-4"
Weight	0.07-0.2 oz.

Our only local hummingbird, the ruby-throat has the distinction of being the bird with the fewest feathers: 940.

[14] While hummingbirds have long been classified in the order Apodiformes along with swifts due to their long wings and extremely short legs, some taxonomists now place hummingbirds in their own Order: Trochiliformes.

Order Columbiformes -- Doves

Family Columbidae

Mourning Dove

(*Zenaide macroura*)

Length	12”
Wingspan	15-18”
Weight	4-6 oz.

Order Passeriformes -- Songbirds/Perching Birds

Family Bombycillidae (Waxwings)

Cedar Waxwing

(*Bombycilla cedrorum****)***

Length	6-7”
Wingspan	9-12”
Weight	1.1 oz.

Flocks (rarely individuals) of cedar waxwings may show up throughout the year, but generally arrive when fruit is ripe on the vines or trees. Waxwings will pick fruit and offer it to mates.

Family Cardinalidae (Cardinals)

Northern Cardinal

(Cardinalis cardinalis)

Length	8-9"
Wingspan	10-12"
Weight	1.2-2.3 oz.

Year-round residents, the cardinals will raise two or more broods in a single season, with juveniles assisting with the feeding of subsequent broods. Juvenile males look like the females. The dominant male in the area will have the brightest colors.

Indigo Bunting

(Passerina cyanea)

Length	4.5-6"
Wingspan	7-9"
Weight	0.5 oz.

Rose-breasted Grosbeak
(Pheucticus lidovicianus)

Length	7-9"
Wingspan	11-13"
Weight	1-2 oz.

Male grosbeaks arrive in April and establish their pecking order before the arrival of the females about a week later.

Family Corvidae (Crows)

American Crow

*(*Corvus *brachyrhynchos)*

Length	16-20″
Wingspan	33-39″
Weight	11-21 oz.

Fish Crow

*(*Corvus *ossifragus)*

Length	14-16″
Wingspan	33-35″
Weight	9-11 oz.

The American crow and fish crow are quite similar in appearance. Fish crow is somewhat smaller and may exhibit a blue sheen in good light. Best distinguished by their call. The Merlin phone app is invaluable for sorting out the calls.

Blue Jay

(Cyanocitta cristata)

Length	9-12″
Wingspan	13-17″
Weight	3 oz.

Family Passerellidae (New World Sparrows)

Dark-eyed Junco
(Junco hyemalis)

Length	5-7”
Wingspan	7-10”
Weight	0.6-1 oz.

Rarely seen in the summer, juncos spend their winters in our area. Common at feeders when safflower seed is offered.

Subspecies
Slate-colored Dark-eyed Junco
Junco hyemalis hyemalis

Eastern Tohee
(Pipilo erythrophthalmus)

Length	7-9”
Wingspan	8-12”
Weight	1-2 oz.

Song Sparrow
(*Melospiza melodia)*

Length	4-7"
Wingspan	7-10"
Weight	0.5-2 oz.

The song sparrow is a year-round resident often found in bushes along the shore of Lake Culpeper.

Swamp Sparrow
(*Melospiza georgiana)*

Length	5-6"
Wingspan	7-8"
Weight	0.5-0.8 oz.

White-throated Sparrow
(Zonotrichia albicollis)

Length	6-7.5"
Wingspan	9"
Weight	0.8-1.1 oz.

White-crowned Sparrow
(Zonotrichia ***leucophrys****)*

Length	6"
Wingspan	9"
Weight	0.9 oz.

Chipping Sparrow
(Spizella passerina)

Length	5-6"
Wingspan	8"
Weight	0.5 oz.

Family Fringillidae (Finches)

American Goldfinch
(*Carduelis tristis*)

Length	4-5"
Wingspan	7.5-9"
Weight	0.3-0.7 oz.

House Finch
(*Haemorhous mexicanus*)
Formerly (*Carpodacus mexicanus*)

Length	5-6"
Wingspan	8-10"
Weight	0.75 oz.

Seen side by side, male house finch and male purple finch in breeding plumage are easy to tell apart. The rest of the year…not so much. Purple finches have shorter tails tend to be pinker than house finches.

Purple finch
(*Haemorphous purpureus*)

Length	5-6"
Wingspan	7.5-9"
Weight	0.6-1.1 oz.

Family Hirubdubudae (Swallows)

Tree Swallow

(*Tachycineta bicolor)*

Length	4.5-5.5"
Wingspan	12-14"
Weight	0.6-0.9 oz.

Northern Rough-winged Swallow

(*Stelgidopteryx serripennis)*

Length	5-6"
Wingspan	11-12"
Weight	0.35-0.55 oz.

Family Icteridae (New World Blackbirds)

Red-winged Blackbird

(*Agelaius Phoeniceus)*

Length	7-9”
Wingspan	12-16”
Weight	1.5-2.3 oz.

Rusty Blackbird

(*Euphagus carolinus)*

Length	8-10
Wingspan	14”
Weight	2 oz.

Once common, the rusty blackbird population has crashed over the past 40 years for reasons not yet fully understood. They occasionally show up around Lake Culpeper in late fall as they migrate south.

Baltimore Oriole

(Icterus galbula)

Length	7-9
Wingspan	9-13"
Weight	0.8-1.5 oz.

Orchard Oriole

(Icteruse spurius)

Length	6-7
Wingspan	10"
Weight	0.6-1 oz.

First year male

Brown-headed Cowbird

(*Molothrus ater)*

Length	6-9
Wingspan	14”
Weight	1-2 oz.

Common Grackle

(*Quiscalus quiscula)*

Length	11-13”
Wingspan	14-18”
Weight	2.5-5 oz.

Family Mimidae (Mimids)

Grey Catbird

(*Dumetella carolinensis*)

Length	8-9”
Wingspan	9-12”
Weight	0.8-2 oz.

Northern Mockingbird

(*Mimus polyglottos*)

Length	8-11”
Wingspan	12-15”
Weight	1.5-2 oz.

Brown Thrasher

(*Toxostoma rufum*)

Length	9-12”
Wingspan	11-13”
Weight	2-3 oz.

Family Paridae (Chickadees and Titmice)

Tufted Titmouse
(*Baeolophus bicolor)*

Length	5.5-6.5"
Wingspan	8-10"
Weight	0.6-0.9 oz.

Nuthatch, titmouse, and chickadees are often seen together. They will eat bugs when bugs are available, do quite well with seeds and berries in the fall and through the winter.

Carolina Chickadee
(*Poecile carolinensis)*

Length	4-5"
Wingspan	6-8"
Weight	0.3-0.4 oz.

Family Parulidae (New World Warblers)

Except for the Yellow-rumped warbler which is present year-round, all the warblers were photographed during spring migration.

Yellow-rumped Warbler
(*Dendroica coronate)*

Length	5-6"
Wingspan	7.5-9.5"
Weight	0.4-0.5 oz.

Northern Parula
(Setophaga americana)
or
Parula
(*Parula americana)*

Length	4-5"
Wingspan	6.7"
Weight	0.2-0.4 oz.

Yellow Warbler

(Parulidae petechia)

Length	4-7"
Wingspan	6-9"
Weight	0.25-0.9 oz.

Cape May Warbler

(Setophaga tigrine)

Length	5-5.5"
Wingspan	7.5-9"
Weight	0.3-0.6 oz.

Black-throated Green Warbler

(Setophaga virens)

Length	4-5"
Wingspan	7-8"
Weight	0.3-0.8 oz.

Chestnut-sided Warbler

(Setophaga pensylvanica)

Length	4-5"
Wingspan	6-8"
Weight	0.3-0.45 oz.

This is a female. The mature male has chestnut-colored flanks which are most pronounced during breeding season.

Palm Warbler

(Setophaga palmarum)

Length	4.7-5.5"
Wingspan	8"
Weight	0.3-0.5 oz.

Prothonotary Warbler

(Protonotaria citrea)

Length	5"
Wingspan	8.75"
Weight	0.45 oz.

Common Yellowthroat

(Geothlypis trichas)

Length	4-5"
Wingspan	6-7.5"
Weight	0.3 oz.

Family Passeridae (Old World Sparrows)

House Sparrow

(Passer domesticus)

Length	6.3"
Wingspan	7.5-10"
Weight	0.85-1.4 oz.

The house sparrow is the world's most abundant wild bird, with a population estimated at 1.6 billion.

Family Regulidae (Kinglets)

Ruby-crowned Kinglet
(Corthylio calendula)

Length	3.5-4.3"
Wingspan	6-7"
Weight	0.2-0.4 oz.

The ruby crown is a feature of the male but is only occasionally visible.

Family Sittidae (Nuthatches)

White-breasted Nuthatch

(Sitta carolinensis)

Length	5-6"
Wingspan	8-11"
Weight	0.6-1 oz.

Family Sturnidae (Starlings)

European Starling

(Sturnus vulgaris)

Length	7.5-9"
Wingspan	12-17"
Weight	2-3.5 oz.

AKA **Common Starling**

Starlings were introduced to the United States as early as the 1870, famously In New York City and Portland, Oregon in the 1890s. They did well, with a U.S. population now over 150 million. Large flocks are known for their mesmerizing murmurations in which flocks continually morph and transform in flight.

Family Troglodytidae (Wrens)

Carolina Wren

(Thryothorus ludovicianus)

Length	5-5.5"
Wingspan	7"
Weight	0.6-0.8 oz

The Carolina wren is a year-round resident and frequent customer at the bird feeder. During breeding season, the male will build several nests; the female will choose her favorite and redecorate it before laying eggs.

The house wren is smaller but more aggressive. It will nest in the eaves and defend its territory against all other birds, often displacing Carolina wrens and chickadees nesting nearby.

House Wren

(Troglodytes aedon)

Length	4-5"
Wingspan	6"
Weight	035-0.4 oz

Family Turdidae (Thrushes)

Eastern Bluebird

(Sialia sialis)

Length	6-8”
Wingspan	10-13”
Weight	1-1.2 oz.

Bluebirds and robins may be found all year in our area, although particularly harsh weather may prod some of them to migrate south.

American Robin

(Turdus migratorius)

Length	9-11”
Wingspan	12-16”
Weight	2.7 oz

Hermit Thrush

(Catharus guttatus)

Length	6-7”
Wingspan	10-12”
Weight	0.6-1.3 oz

Uncommonly seen, the hermit thrush is most often in the underbrush searching for food. These thrushes will have migrated by November, returning in April.

Family Tyrannidae (Tyrant flycatchers)

Eastern Phoebe

(Sayornis phoebe)

Length	5.5-6.5”
Wingspan	10-11”
Weight	0.6-0.7 oz

The phoebe and pewee are often hard to tell apart. Distinct wing bars and a pointed crest are good indicators of the pewee. The phoebe is more likely to be bobbing its tail.

Eastern Wood-Pewee

(Contopus virens)

Length	5-6”
Wingspan	9-10”
Weight	0.5 oz

Eastern Kingbird
(Tyrannus tyrannus)

Length	7.5-9"
Wingspan	13-15"
Weight	1-2 oz

The kingbird is our largest and most common flycatcher, easily distinguished by the white tip of its tail feathers. It perches openly on treetops and power lines. The great crested flycatcher is much less common and more furtive.

Great Crested Flycatcher
(Myiarchus crinitus)

Length	7-8"
Wingspan	13"
Weight	1-1.4 oz

Family Vireonidae (Vireos)

White-eyed Vireo

(Vireo griseus)

Length	4-5"
Wingspan	6.7"
Weight	0.3-0.5 oz

The vireo tends to stay in the woodland where it feeds of insects among the tree branches.

References

Print

American Bird Conservancy, *Wild Bird Longevity Records*

Bird, D.M., *The Bird Almanac*, Firefly Books, Buffalo, N.Y. 2004

Bowles, H.E, Evaluating and Treating the Reproductive System, *Clinical Avian Medicine, Vol II*

Brinkley, E.S., *National Wildlife Foundation Field Guide to Birds of North America,* Stirling Publishing, New York 2007

Callaghan, C.T. et al, Global abundance estimates for 9,700 bird species*, PNAS* 2021 vol.118 No.21

Parisi, G, *In a Flight of Starlings*, Mondadon Libri, Milan 2021

Rosenberg, K.V. et al, Decline of the North American Avifauna, *Science*, AAAS, 2019

Stark, J.M., Phenotypic flexibility of the avian gizzard: rapid, reversible and repeated changes of organ size in response to changes in dietary fibre content, *Journal of Experimental Biology, Volume 202, Issue 22* November 1999

Wasser, D.E. and Sherman, P.W., Avian longevities and their interpretation under evolutionary theories of senescence*, Journal of Zoology*, 25 January 2010

Yong, E*., An Immense World*, Random House, New York 2022

Internet

INaturalist (https://www.inaturalist.org/) - For a comprehensive view of local animals and plants go to https://www.inaturalist.org/projects/biota-of-culpeper-county-vaiNaturalist

Cornell Labs (https://www.allaboutbirds.org/)

Phone

The Merlin app will identify birds by listening to their vocalizations.

INaturalist also has a phone app. Post pictures from your phone and iNat will identify them with a good degree of accuracy.

	Average Lifespan (years)	Maximum Documented (wild)	Maximum Lifespan (captive)
American Crow	7	16	
American Goldfinch	5	11	
American Robin	2	14	13
American wigeon	2	21	
Bald Eagle	20	38	
Baltimore oriole		12	
Barred owl		18	
Black-capped Chickadee	4	12	
Black-throated green warbler		6	
Blue Jay	7	17	26
Brown thrasher		13	12
Canada Goose	10	28	33
Cape May warbler		4	
Carolina Chickadee	2	11	
Chestnut-sided warbler	2	7	
Chipping Sparrow	2	10	
Cooper's hawk		12	
Double-crested cormorant		18	
Downy Woodpecker		11	
Eastern Bluebird		10	
Eastern towhee		12	
Eastern screech owl		14	
Eastern wood pewee		7	
European Starling		15	
Great Blue Heron		23	
Great egret		23	
Green Heron		7	
Hairy Woodpecker		16	
Hermit thrush		9	
House Finch	5	11	
House Sparrow		13	19
House Wren		9	
Hummingbird	9	12	

Lifespan of selected local

	Average Lifespan (years)	Maximum Documented (wild)	Maximum Lifespan (captive)
Indigo bunting		11	
Killdeer		11	
Least sandpiper		16	
Lesser scaup		18	
Lesser yellowlegs		5	
Mallard	5	27	29
Mourning Dove	5	19	30
Mute Swan	15	27	28
Northern Cardinal	3	15	
Northern Flicker	6	9	
Northern Mockingbird		14	20
Orchard oriole		10	
Osprey		26	
Pectoral sandpiper		6	
Prothonotary warbler	2	11	
Purple Finch	5	12	
Red-bellied woodpecker		12	9
Red-headed woodpecker		10	
Red-shouldered hawk		20	
Red-tailed Hawk	12	30	
Red-winged Blackbird	2	15	
Ruby-throated Hummingbird	9	12	27
Song Sparrow		11	
Tree Swallow		12	
Trumpeter swan		23	32
Tundra swan		24	11
Tufted Titmouse		13	
Turkey Vulture	16	37	
White-breasted Nuthatch		10	
White-eyed vireo		7	
White-throated Sparrow		10	14
Yellow Warbler	3	11	
Yellow-rumped warbler		7	

Lifespan of selected local birds

Also by the Author

Wazzat Bug?

Wazzat Beetle?

Wazzat Lep?

The Ant Farm

Hemolymph Moon

Swatting Gnats

There's Poo on My Shoe

Amoebae Are Not Self-Aware

The Zoophabet

Thirty Years in Flight

Charlie Gets Respect

All available at Amazon.com

Made in the USA
Monee, IL
05 November 2026

67985485R00055